For Griffin, Chase, & Haley—our planet is a much
better place with the three of you.
—Frank

For Madison and Barrington—may you both have many more trips
around the sun and enjoy the Earth's beautiful wonders.
—Charnaie

For Peter, you are my world.
—Kayla

---

Just like it takes all of us to work as a global team to help nourish and grow our
Earth, a book takes a team too. We are grateful to the editing team, led by Sarah
Rockett, and the art team, led by Melinda Millward, for their guidance and creativity
in creating *A Planet Like Ours*. The whole team at Sleeping Bear Press is so talented
and we are grateful to them for the opportunity to create this book. And, of course,
we are grateful for Kayla Harren and her stunningly beautiful and unforgettable art.

—Frank and Charnaie

---

Text Copyright © 2022 Frank Murphy and Charnaie Gordon
Illustration Copyright © 2022 Kayla Harren
Design Copyright © 2022 Sleeping Bear Press

## SLEEPING BEAR PRESS™

2395 South Huron Parkway, Suite 200, Ann Arbor, MI 48104
www.sleepingbearpress.com
© Sleeping Bear Press
Printed and bound in the United States.
10 9 8 7 6 5 4 3 2 1

Library of Congress Cataloging-in-Publication Data

Names: Murphy, Frank, 1966- author. | Gordon, Charnaie, author. | Harren, Kayla, illustrator.
Title: A planet like ours / by Frank Murphy and Charnaie Gordon; illustrated by Kayla Harren.
Description: Ann Arbor, MI : Sleeping Bear Press, [2022] | Audience: Ages
4-8 | Summary: "Our planet Earth is as individual and special as each one of us. It's ability to sustain
and nurture life is unique in our solar system—and beyond. In this book, celebrate all the wonderful
qualities of our Earth while learning how to protect her for future generations"– Provided by publisher.
Identifiers: LCCN 2022006571 | ISBN 9781534111530 (hardcover)
Subjects: LCSH: Environmental protection–Juvenile literature. | Earth (Planet)–Juvenile literature.
Classification: LCC TD170.15 .M87 2022 | DDC 363.7–dc23/eng/20220324
LC record available at https://lccn.loc.gov/2022006571

# A PLANET like OURS

By Frank Murphy and Charnaie Gordon
Illustrated by Kayla Harren

There are hundreds
and hundreds of billions

of planets in the universe.

Maybe more.

Maybe too many to count.

But there's only one planet like ours.

And we need a planet like ours
to nourish us, protect us, and give us life.
A place to call home.

Our planet. Our Earth.

We don't always take care of our planet.

Sometimes, we take too much.

Sometimes, we are careless.

We need to care for those
who call Earth home.
We must.

If not us, then who?

# Our planet. Our soil.

It starts far beneath our feet. Under the ground there is water.
Groundwater supplies wells and springs with fresh water.

It gives us the
water we drink.

It helps to grow
our food.

Pollution and chemicals can ruin our soil and our groundwater.

## Let's protect our soil.

# Our planet. Our water.

Earth is special. Water is everywhere!

Without water there would be no plants...

no animals...no people!

So we must take care of lakes, rivers, and oceans.

Litter ends up in our waterways and seas.
And plastic never goes away.
Bit by bit—it all adds up.

Let's protect our water.

# Our planet. Our animals.

From earthworms to elephants and bumblebee to bison,
our planet is home to about 10 million different species.

And they all work together to keep
our ecosystem balanced.
We're all connected.

Taking care of creatures means taking care of their habitats.
## Let's protect all living things.

# Our planet. Our trees.

Our trees help keep the atmosphere cool.

They grow branches that grow leaves that give us shade.

Our trees grow roots that hold tight underground and prevent erosion.

They give us foods like apples and oranges, pomegranates and pecans.

Trees are an important resource.
They can grow from tiny seeds into giant, towering trees.
A tree can grow...and grow...and grow for hundreds of years!

Let's protect our trees.
If not us, then who?

# Our planet. Our air.

When our air gets polluted, it hurts all living things.

We all need air to live.

Earth's atmosphere needs clean air.

Pollution comes from factories and energy power plants.

Cars and planes.

Chemicals and fumes
from spray cans.

But we can all help make a difference. Every little bit counts.
**Let's protect our air.**

# Our planet. Our people.
Our cities, towns, and neighborhoods.

Where else would we
toss a ball, together?

Push a swing, together?

Splash in the ocean, together?

We build communities, together.

We learn new things, together.

And we make changes, together.
We need to care for each other.

Let's protect our planet by protecting each other.

Because remember,
we all need a planet.

A planet that nourishes and protects us.
That gives us life.

Our soil, our water,
our animals and plants and trees,
our air, our atmosphere...
and each other.

A planet just like ours.

## Note from the Authors and Illustrator

We wrote and illustrated *A Planet Like Ours* as a love letter to Earth—and to you, the people who will need to help protect our planet. The problems here on Earth are ones you've inherited, but without new, young leaders being proactive, our planet is in danger. We hope *A Planet Like Ours* inspires you to research more about Earth to learn about its vulnerabilities and strengths. We hope it inspires you to act with passion and great intentions to protect and help nourish our planet. It might be small acts like turning off a light, picking up litter, riding a bike to school, using a reusable tote bag instead of a plastic one. Or big acts like writing to a congressperson, planting a tree that will live for decades, participating in a community garden, or one day pursuing a career that strives to protect our Earth. The messages and art in this book do not cover every aspect of our environment, but we tried to touch upon many. We did make sure to end with one of the most important elements of saving our planet—each other. We need to nourish and protect each other, just as our Earth nourishes and protects us.

Earth has always been here for us. Each one of us—each different identity, generation, culture, ethnicity, or race; every aspiration, wish, and hope, in every geographic location. We must do better for Earth and for each other. We all share the same home, a home for all of us—our planet. Our Earth. If not us, then who?

—Frank, Charnaie, and Kayla

# Activities

- Start an Earth Care journal! Use your journal to tak notes while completing the activities below and to reflect upon what you've learned. Write and sketch observations about the Earth around you. And use it to think on new ways you might be able to help the environment and Earth's people. Just grab a notebook and get drawing and writing!

- Save fresh water! Did you know that 71 percent of Earth's surface is water? But only 3 percent of that is fresh water. And only a fraction of that is available for us to use. This is why it's so important to conserve our fresh water. Try this:
  1. Before brushing your teeth, place a bowl under the faucet.
  2. Brush your teeth for two minutes, leaving the water running the whole time.
  3. Use a ruler to measure the depth of water in the bowl.
  4. The next time you brush your teeth, place the same bowl under the faucet. This time, only turn the water on when necessary.
  5. Measure the depth of water. Compare the two amounts.
  6. Take it further and figure out how many times you brush your teeth each day, then multiply that amount by 365 days. Multiply that number by both amounts. You'll be surprised by the difference between the two numbers—and by how much water you can save each year!

  NOTE: Use your collected water to give thirsty plants a drink, and only complete this activity once. Doing it more will waste precious fresh water.

- Reduce your paper product use! Did you know that people use more than 13 billion pounds of paper towels every year—in America alone? Try these steps to reduce your paper use at home:
  1. Challenge your family to see how long they can make a roll of paper towels last.
  2. Use reusable dish towels or washcloths instead.
  3. Switch to cloth napkins for mealtimes. You can often find these at resale stores or make your own with cotton fabric!
  4. Ask your family if you can switch to paper towels made from a renewable source like bamboo.